Ivamauro Ailton de Sousa Silva
Thyago Costa

# Atmospheric changes and the occurrence of diseases in Cáceres

Ivamauro Ailton de Sousa Silva
Thyago Costa

# Atmospheric changes and the occurrence of diseases in Cáceres

Mato Grosso - Brazil

ScienciaScripts

**Imprint**



Cover image: www.ingimage.com

This book is a translation from the original published under ISBN 978-3-330-76479-8.

Publisher:
Sciencia Scripts
is a trademark of
Dodo Books Indian Ocean Ltd. and OmniScriptum S.R.L publishing group

120 High Road, East Finchley, London, N2 9ED, United Kingdom
Str. Armeneasca 28/1, office 1, Chisinau MD-2012, Republic of Moldova, Europe
Managing Directors: Ieva Konstantinova, Victoria Ursu
info@omniscriptum.com

Printed at: see last page
**ISBN: 978-620-8-50339-0**

# SUMMARY

## INTRODUCTION

Climate and its variations are important elements in shaping geographical space, in the composition and transformation of the environment, in the availability of natural resources and in the socio-economic characteristics of urban centres (Souza, 2007).

The relationship between the atmospheric environment and human beings is an eminently interdisciplinary area of study that has interested specialists from various fields: doctors, architects and meteorologists.

Although its nature is appropriate for a geographical approach, Mesquita (2005) and Sales (2006) point out that few geographers, especially in Brazil, have dedicated themselves to the subject, due, among other reasons, to the complex relationship and the almost deterministic aspect of this subject.

Many of the "winter" illnesses, which correspond to the months of "June, July and August", are directly or indirectly related to climatic conditions, as temperature anomalies, pollution or relative humidity can lead to a series of urban problems that compromise the quality of the population.

Confaloniere (2003) points out that two factors are predominant in the influence of climate on health: the first relates to the regular succession of climates that influence biological behaviour and can culminate, for example, in cases of respiratory diseases or facilitate recovery from them.

The second refers to extreme climatic/meteorological phenomena with the possibility of accidents and trauma to those involved, such as victims of floods or prolonged droughts. Therefore, eccentric climatic conditions cause problems in the clinical picture of health, which is the result of the reactions and vulnerabilities of the human organism.

In this context, Souza's research (2007) found that during periods of prolonged drought, temperature fluctuations and relative humidity mostly below 60 per cent, there was an increase in the number of cases of hospitalisation for respiratory problems.

In this way, it is possible to emphasise that climatic conditions compromise the population's quality of life, caused by respiratory problems, but Castilho (2006) points out that each individual, as well as each social group, has singularities in their living conditions and, consequently, different exposures to social vulnerability.

From this perspective, the climate has become relevant in studies dealing with the quality of life of the population, particularly in urban centres, since there is a strong relationship between climatic characteristics and the incidence of some diseases in tropical environments and this has an immediate impact on quality of life.

One of the socio-urban factors that most affect the population is the thermal discomfort caused by the increase in temperature/heat accumulation within the core of urban centres. According to Alvares Júnior *et. al.* (2002, p. 131) "education, food, hygiene, basic sanitation, medical expenses and social relations have a great influence on the health of the population, but they also play a role in determining the incidence and spread of an illness.

In addition to these aspects, Pitton (2007) states that climate or atmospheric conditions are factors of great significance, the importance of which varies according to the disease in question and the physical, psychological and cultural characteristics of individuals.

The aim of this research is to present the relationship between climate and health, showing that climatic factors such as lack of rainfall, low relative humidity, increased pollution, characteristic of the dry season, successive days of rain intrinsically affect the clinical picture of health. The study was carried out in the city of Cáceres, where the elements and factors that imply the incidence and potentialisation of illnesses will be addressed, using the methodology called "climate-environment perception", proposed by Sartori (2000).

The climatic conditions of the city of Cáceres are characterised by different variations due to its tropical continental climate, with two well defined seasons throughout the year. The rainy season, for example, takes place between November and April, while the dry season is distributed between the months of May and October, with low relative humidity values, making it a very worrying time for human health. According to the National Meteorological Institute (INMET), a state of alert is defined when the humidity is between 20% and 30%.

From this perspective, the research problem is based on the following questions: I. Under what circumstances do climatic elements (heat, temperature, drought, relative humidity) interfere in the genesis and potentialisation of illnesses in Cáceres? II. In what way do health professionals and the population of Cáceres interpret the influence and relationship of the climate on the clinical picture of health and even aggravate some health problems.

The aim of this work is to investigate the influence of climatic factors on the genesis and potentialisation of diseases, indicating the most critical periods and the most common diseases associated with drought and exceptional atmospheric changes. In addition, the aim is to:

- Characterise the climatic aspects of the area under study;
- Emphasise the relationship between climate and the incidence of disease;
- Identify the periods that offer the greatest risk for the emergence of diseases that jeopardise the health of the population;
- Organise a detailed description of the most common diseases in the two seasons;

It is in this context that it is extremely important to study climate and the variation of atmospheric elements in the emergence and worsening of illnesses, as well as in health, since the human body reacts to the climatic conditions imposed on it by the environment. This research therefore set out to understand the relationship between the climatic conditions of the dry season and their implications for health in Cáceres.

The research is structured in 4 (four) distinct and complementary parts. Firstly, the theoretical framework is presented, which guided the research, addressing conjectures inherent to climate and health/illnesses, as well as themes such as climate perception and vulnerability.

The second part, on the other hand, looks at the research process, explaining the methodological aspects as well as the procedures and operational stages of the work in detail. The third part of the book presents a brief characterisation of the study area, mainly highlighting its climatic characteristics.

The results of the research are discussed at the end of the book, which highlights the climate perception of some health professionals and the participation of civil society about the influence of climatic conditions on the incidence of illness.

The final considerations include an extension of the research and a general exposition of the subject, the problems identified in the research that were answered in terms of interpreting the possible relationships between climate and health. This section also highlights recommendations for reducing the effects of exceptional climatic and atmospheric conditions.

Ultimately, we can collaborate with information that can be used for environmental and urban planning, since the well-being of the population is closely linked to climatic conditions and especially the environmental situation in which they live.

# CHAPTER 1

## Climate and health: atmospheric changes and the incidence disease

Throughout history, man has felt the effects of atmospheric conditions, such as the slow fluctuations of the climate, which cause migrations, the extremes of seasonal weather, which cause famine, and the various disasters, which lead to death and malnutrition" (PITTON and DOMINGOS, 2004, p.77).

According to Souza (2007), geography's contribution to analysing the rhythm and variability of climate in health studies dates back to the beginning of the 20th century. Maximilien Sorre (1880-1962), who produced both a single and diversified work, sought to discuss the experience of man and his relationship with the Earth, correlating it with biology, medicine, sociology and geography.

In Brazil, the work of Afrânio Peixoto (1938) was one of the pioneers in establishing correlations between certain diseases and the country's climatic conditions. Introducing Medical Geography studies, it provides a detailed explanation of the manifestation of numerous diseases and their correlation with the natural environment, demythologising and criticising climatic or tropical diseases.

Ayoade (1986), for his part, emphasised that the influence of climate on human health is both direct and indirect, whether in a harmful or beneficial way. For this author, thermal and hygrometric extremes accentuate the body's weakness in the fight against illness, intensifying inflammatory processes and, consequently, creating contagious conditions.

On the contrary, fresh air (with a mild temperature, humidity and moderate radiation) has therapeutic properties. However, temperature can be the main trigger for some types of illness, more than any other climatic element, such as infant mortality from respiratory illnesses and acute respiratory infections:

> "[...] climate and/or atmospheric conditions are factors of great significance, the importance of which varies according to the disease in question and the physical, psychological and cultural characteristics of individuals" (PITTON and DOMINGOS, 2004, p.76).

Thus, the effects of climate and atmospheric weather on socio-environmental quality are still not well understood. However, there are a considerable number of studies that show that climatic conditions influence biological rhythms, which interfere in all human activities and functions. We can mention theoretical contributions developed by Pitton (2002), Zem (2004), Barros (2006), Castilho (2006), Souza (2007) and others.

Confaloniere (2003) points out the climate/health relationship by reporting two predominant factors in the influence of climate on health: one relating to the regular succession of the climate with an influence on biological behaviour; the second relating to extreme climatic/meteorological phenomena with the possibility of accidents and trauma to those involved. The first could culminate, for example, in cases of respiratory diseases or facilitate recovery from them. The second would be the case of victims of floods or prolonged droughts.

Mariano (2010) emphasises the importance of understanding climate parameters in order to understand the relationship between climate, health and the population:

> climatic parameters (air temperature, relative humidity, precipitation, atmospheric pressure and winds) have a direct and indirect influence on human health, since society is in permanent contact with the environment. The atmosphere-surface interface, for example, enables thermal, water and gas exchange (MARIANO, 2010, p.1).

Silva and Mendes (2012, p.127-128) point out the importance of geographically analysing the climate in relation to types of respiratory diseases, believing that in environments such as "the Cerrado there are climatic conditions marked by continentality, which are conducive to worsening respiratory problems. Especially during the dry season, when the relative humidity reaches extremely low levels and fires occur more frequently, further compromising air quality".

However, Silva and Mendes (2012) point out that the influence of climate on health depends not only on climatic elements, but also on the environment in which the population lives (urban or rural). In this sense, the authors highlight the environmental and microclimatic differences between urbanised areas and rural areas:

> the difference means that, not infrequently, the pollution generated in the countryside is concentrated on the urban perimeter, where the majority of the municipality's population is located, increasing respiratory problems. This factor possibly indicates what happens in the city of Patrocínio in the aforementioned months, when the winds overcome the calm and end up bringing the harmful effects of fires and dust into the urban perimeter (SILVA and MENDES, 2012, p. 132).

Oliveira (2014) states that air temperature and relative humidity are the climatic parameters that most affect human health, causing concerns and illnesses such as pneumonia, bronchitis, asthma and chronic tonsil diseases. According to Oliveira (2014), the months with the most respiratory problems are March, April, May, June and July. For Oliveira (2014), this is the period when the autumn-winter season transitions, which causes significant fluctuations in temperature, relative humidity and changes in rainfall.

Domingos and Pitton (2004) state that among the various factors that play a role in determining and spreading an illness, climate and atmospheric conditions stand out, the intensity of which will depend on the illness in question and the physical, psychological and cultural characteristics of each individual, which will influence the extent of their susceptibility to developing a particular illness. The authors add that climate elements can affect health both directly and indirectly:

> Climatic parameters such as air temperature, humidity, precipitation, atmospheric pressure and winds affect health directly (feelings of comfort, mortality and morbidity from systemic diseases) and indirectly (infectious diseases carried by vectors: air, water, soil and food), as the human body is in permanent contact with its atmospheric environment through thermal, water and gas exchange (DOMINGOS and PITTON, 2004, p. 77-78).

Oliveira (2014) states that the biggest victims of health problems associated with climate change are children and the elderly due to their fragility or lack of resistance. Barros (2006) explains that these groups are more vulnerable because their immune systems are susceptible to sudden changes in temperature and humidity.

Barros (2006, p. 33) states that "the change in temperature, the variation in humidity, the increase in the vehicle fleet and exposure to harmful particles are factors that interfere with health and contribute to an increase in risks, especially for children and the elderly".

Souza (2007) shows that the behaviour of climatic parameters such as air temperature, relative humidity and dry days have a natural effect on the incidence of hypertensive crises in people over 40, regardless of gender.

According to Barros (2006, p. 34), cold and heat have other influences: heat, for example, favours the multiplication of germs, reduces gastric secretion, decreases organic defence reactions and greatly favours intestinal infectious diseases. Cold can have some acute local effects and aggravate certain illnesses.

Barros (2006) states that the weather and climate are not exclusively responsible for this. In interaction with the other elements of the geographical space and according to the susceptibility of individuals, they increase environmental risks, making it possible to aggravate existing diseases.

According to Castilho (2006), climate is not in itself a risk factor for the onset of these diseases, but it is one of the elements influencing environmental risk factors (temperature and ultraviolet radiation, among others) and lifestyle risk factors (diet and obesity, physical exercise, smoking, cholesterol levels, coagulation factors and susceptibility).

Considering climatic aspects and their implications for increased morbidity from diseases/illnesses, studies carried out by Domingos and Pitton (2004); Castilho (2006), Barros (2006) and Souza (2007) found a seasonal trend in illnesses, which are more frequent in periods with a prevalence of low relative humidity, a range of temperatures, sharp thermal extremes and a predominance of droughts or irregular droughts.

According to Souza (2007) many diseases occur seasonally, which is probably related to climatic conditions, such as scarlet fever, diphtheria and jaundice, which in Switzerland occur mainly in winter, while measles, flu and chickenpox are more common in spring.

According to Pitton and Domingos (2004), adverse situations, such as heatwaves in summer and cold snaps in winter, affect health and well-being in various ways. The combination of low temperatures and wind can make the air temperature noticeably colder and can easily lead to hypothermia (body temperature below 35°C), which is produced by the stress of excessive cold (thermal sensation).

The heart rate slows down, breathing becomes slower and blood vessels contract (increasing blood pressure), which can lead to loss of consciousness (fainting), freezing of the extremities and cardiac arrest. Pitton and Domingos (2004) state that:

> "as some symptoms are linked to changes in the weather and these can be predicted by meteorological services, having a certain number of regional and local studies that indicate the meteorological situation that determines the development of certain diseases, there would be the possibility of warning the population and asking them to take the necessary initiatives and precautions" (PITTON and DOMINGOS, 2004, p.84).

The two main indicators used to analyse collective health conditions are the number of hospitalisations for each disease (morbidity) and mortality **(SOUZA and SANT'ANNA NETO, 2006).** In this context, Confalonieri (2003) states that reducing the impacts caused by climate variability on the Brazilian population can only be done by understanding and modifying the social vulnerability factors that affect these populations in their specific geographical contexts.

## 1.1 Environmental and climate perception: notions of weather and climate

In order to understand the perception of society, we first need to understand the greater whole of which we are a part as a set of physical events, and man himself as a physiological structure.

Perception is both the response of the senses to external stimuli and the purposeful activity in which certain phenomena are clearly registered, while others recede into the shadows or are blocked. According to Sartori (2000), environmental perception should be understood as society's response to stimuli from the environment in which it lives.

Thus, Faggionato (2005) explains that environmental perception has been defined as "man's awareness of the environment" and its study is of fundamental importance if we are to better understand the interrelationships between society and the environment, their expectations, satisfactions and dissatisfactions, judgements and behaviour.

In this way, it is also a process of environmental cognition, which is the direct perception of the individual, selectively acquiring information from the environment (SOUZA, 1998). According to Sartori (2000):

> Climate perception has two focuses: one, the perception of time, observing knowledge of the more rhythmic question of how meteorological time evolves over chronological time; the other, psycho-physiological perception, i.e. each individual will react differently to changes in weather and climate.

According to Sartori (2000), meteorological data, when analysed from a strictly statistical point of view, often camouflages its repercussions on lived space, since the most exceptional episodes in climate perception are not always those that have the greatest repercussions on space.

Barros (1998) analyses climate perception and considers it essential to recognise the reaction that a given stimulus causes, i.e. you need to know what sensation an object or climate phenomenon/event will cause. Perception therefore encompasses sensation.

For Sartori (2000), human experience develops from sensations, at first, and perception in a continuous act, as the brain interprets the stimuli received. Through physiological and behavioural adjustments, man is remarkably adaptable to his environment.

According to Barros (2006), cyclical climate change influences biological rhythms, which affect all human activities and functions. However, human beings show very large individual variations in their adaptability, which interferes with their greater or lesser sensitivity to weather and climate, and thus their comfort and health.

Thus, some people will feel more than others, and these people can also be considered time-sensitive, i.e. they show "psychophysiological reactions induced by the type of weather that occurs on a certain day or on two or three previous days".

Another perceptual issue is knowing how to differentiate between two climatological concepts: climate and weather. There is often confusion between what weather is and what climate is. In fact, these two elements are interrelated, since climate can be described simply as the average temperature, rainfall and wind observed over a given period of time, which varies from months to millions of years (usually 30 years) (Barros, 2006).

In our daily lives, the media announces the weather forecast, which is an estimate of what is expected to happen in terms of temperature and rainfall in a short space of time (usually a week). In this sense, the weather is constantly changing, so that one day it can be hot and sunny, and the next day it can be much colder, raining and very windy. Climate is different in that it doesn't change as frequently as weather. In order to define it more precisely, it is necessary to consider the average weather over a long period of time, to avoid seasonal anomalies (AYODE,2007).

Another common confusion, according to Sartori (2000), is to think that any atypical or extreme event is the result of climate change, such as the occurrence of a very cold winter. There have always been extremes of cold and heat, regardless of climate change. What is projected, however, is that climate change will affect the frequency and intensity of anomalies or extremes. It is only when the average weather, in time and space, is calculated that it becomes clear that the planet has variations and variability.

In this context, weather and climate represent combinations realised in the atmosphere by certain respective values of temperature, humidity, pressure, wind, electric charge, etc. They are the states of the atmosphere. But weather is nothing more than a passing, accidental combination while climate is a set of stable trends that result from permanent conditions over a long period (PÉDELABORDE, 1991).

According to Martins and Teixeira et. al (2007), climate is one of the elements of the environment that most influences the adaptation of living beings in the various existing environments. An interesting fact is its influence on the behaviour and well-being of living beings, especially humans. In places with a hot climate, it is associated with irritability, certain emotional outbursts and even a possible state of stress in the population.

## 1.2 Elements of climate and types of drought

Ayoade (2007, p. 2) defines climate as "characteristics of the atmosphere, inferred from continuous observations over a long period, climate deals with a larger number of data being variability, extreme conditions analysed up to 35 years".

Therefore, a more in-depth analysis is needed to understand the climate of a particular region. Climate is about generalisation across a range of data, while weather is about specific, ephemeral, short-lived and frequently changing events (AYOADE, 2007).

The great variation in the planet's climate is the result of the interaction of climatic factors, which are responsible for the Earth's great climatic heterogeneity and are directly related to the geography of each part of the Earth's surface (AYODE, 2007).

According to Borsato (2016), climate elements are important concepts and categories for defining the climatic conditions of a particular place, region or continent. Conceptually, the elements of climate are the basic attributes for defining the climatic type of a given region, considering mainly: solar radiation, atmospheric pressure, temperature, cloudiness, winds, air masses, precipitation and air humidity.

In this context, all these elements are connected and form part of the Earth's climate system, providing different variations and conditions. For clinical health, according to Confaloniere (2003) and Mariano et *al* (2010), some are more important and can be related to the onset of illnesses and even their aggravation.

Climatic drought, for example, presents atmospheric conditions that cause negative effects on health and also risks to the population. Climatic drought is defined by insufficient rainfall, which classifies it as a drought, and therefore poor distribution (BRASIL, 2004).

Carvalho (2006) states that it is necessary to understand some concepts of climatology, since the word drought is often used incorrectly and confused with aridity. The author defines aridity "as a permanent climatic characteristic", while meteorological drought "is an extreme process that occurs in a given area and place, divided into four types": permanent, seasonal, irregular/variable and invisible drought (Table 1).

Table 1 - Types of drought and their predominant characteristics

| Type of drought | Features |
|---|---|
| Permanent | It occurs in desert climates, where the vegetation has adapted to the aridity and there are no watercourses. These only appear after the rains, which are usually very heavy storms. This type of drought makes agriculture impossible without permanent irrigation |
| Seasonal | Seasonal drought is a particular feature of regions with a semi-arid climate. In these regions, vegetation reproduces because adapted plants generate seeds and then die, or remain dormant during the drought. In these regions, rivers only survive if their water comes from other regions where the climate is humid. This type of drought makes it possible to plant crops as long as it's rainy, or by irrigation |
| Irregular | Irregular droughts can occur in any region where the climate is humid or sub-humid. These are droughts whose period is short and uncertain. They are usually limited in area, not in large regions, do not occur in a defined season and there is no predictability of their occurrence. It is believed that the season in which they are most common is summer, as there is an increase in evapotranspiration due to the increase in sunshine |
| Invisible | Invisible drought: This type of drought is the worst of all, because precipitation is not interrupted, but the rate of evapotranspiration is greater than the rainfall rate, causing an imbalance in regional humidity. This imbalance generates a reduction in air humidity, which in turn increases the rate of evapotranspiration, which in turn feeds the loss of underground moisture to the atmosphere, which returns it in the form of rain, but not enough to increase soil humidity |

Elaboration: Thyago da Costa, 2017 Source: Carvalho (2006)

The Brazilian territory began to incorporate the categories of dry areas, aridity and drought indices quite recently, in the context of the delimitation of the Semi-arid Northeast (BRASIL, 2004). For the tropical climate, which is the predominant type in the area under study, the term "irregular drought" will be used throughout the paper to refer to the period of drought that occurs between the months of May and October.

Otamar de Carvalho (2006) emphasises that droughts (seasonal and irregular)

interfere with economic dynamics (industries and agriculture), but also have an impact on social activities (consumption, supplies and health).

## 1.3 The relationship between climate conditions and society: impacts and vulnerability

A review of various published works points to the possible harmful effects of climatic factors on human living conditions. Souza (2007) states that it is possible to find several studies that show how cyclical climate changes influence the biological rhythm of human beings, significantly interfering in their activities and functions.

According to Pitton and Domingos (2004, p. 77) throughout history, man has felt the effects of atmospheric conditions, such as the slow fluctuations of the climate, which cause migrations, the extremes of seasonal weather, which cause famine, and the various disasters, which lead to death and malnutrition".

Faced with this problem involving climatic conditions and health, Souza (2007) has drawn up a very clear flowchart that allows us to visualise the main potential and effects of climate on human health (Figure 1).

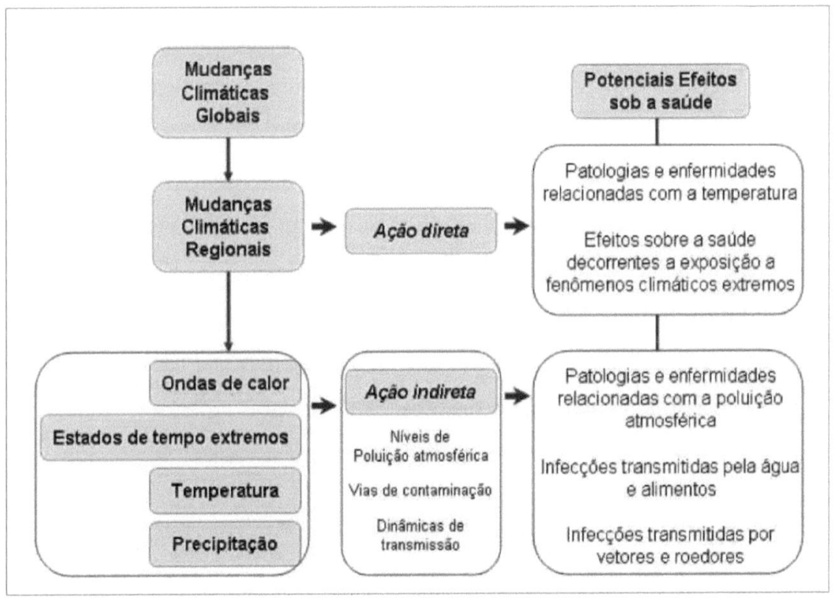

Figure 1 - Flowchart: potential effects of climate change on human health
Source: SOUZA, 2007

According to Fonseca (2004), health is directly linked to the environment (understood as the inseparable interaction between society and nature), since the conditions and/or changes in the natural environment are only important to humans when they are perceived or affect their well-being and way of life. And the climate, due to its cyclical changes and unexpected and damaging variations for man and the social environment (in general), is certainly a factor that interacts directly with human health.

For Mesquita (2005), the search for relationships between climate and health needs to be stimulated. The relationships of this interaction need to be studied, focusing on the multiplicity of environmental aspects and factors involved. The debates and environmental concerns bring up themes such as sustainability and quality of life,

It therefore needs serious and supported discussions in search of data and, above all, action.

In turn, Sobral (1988) and Barros (2006) point out that one of the main approaches to studies related to climate and respiratory diseases are epidemiological studies, which focus on the prevalence of diseases in populations exposed to different levels of relative humidity, wind, temperature and air quality.

The relationship between climate and health has also been proven by Abreu and Ferreira (1999). The authors state that prior knowledge of atmospheric conditions can help prevent or reduce the occurrence of certain diseases.

According to Souza (2007), stable weather conditions are unfavourable for the dispersion of pollutants in the atmosphere, such as weak winds and lulls, low relative humidity and a lack of precipitation. On the other hand, unstable types of weather, such as frontal systems, provide a favourable environment, with ventilation and precipitation, facilitating the dispersion of pollutants.

Thus, rain plays a fundamental and very important role, "washing" the atmosphere and considerably reducing the levels of contaminants, especially suspended particulate matter, since pollution occurs in a cumulative process (SOUZA, 2007, p. 84).

In short, there are various natural processes for purifying the environment and the atmosphere, but they are currently not enough to "clean up" the large amount of anthropogenic substances that exist. The extreme of this problem is in large cities (and also in medium-sized cities), where a large amount of substances are emitted into the atmosphere through various activities and the large concentration of people.

According to Confalonieri (2003, p. 2000) "the study of the social and environmental vulnerability of populations subject to the effects of climatic impacts on their physical integrity and well-being is of fundamental importance for guiding preventive actions".

Vulnerability to climatic conditions, according to Confalonieri (2003), can be defined as the product of physical exposure to a natural hazard and the ability to prepare for and recover from the negative impacts of a disaster, as well as the characteristics of a group, or even a person, in being able to anticipate, resist and resolve the impacts, which may be aggravated by the influence of the climate.

According to Confalonieri (2003, p. 203), "reducing the impacts caused by climate variability on the Brazilian population can only be achieved by understanding and modifying the socio-environmental vulnerability factors that affect these populations in their specific geographical contexts" (Figure 2).

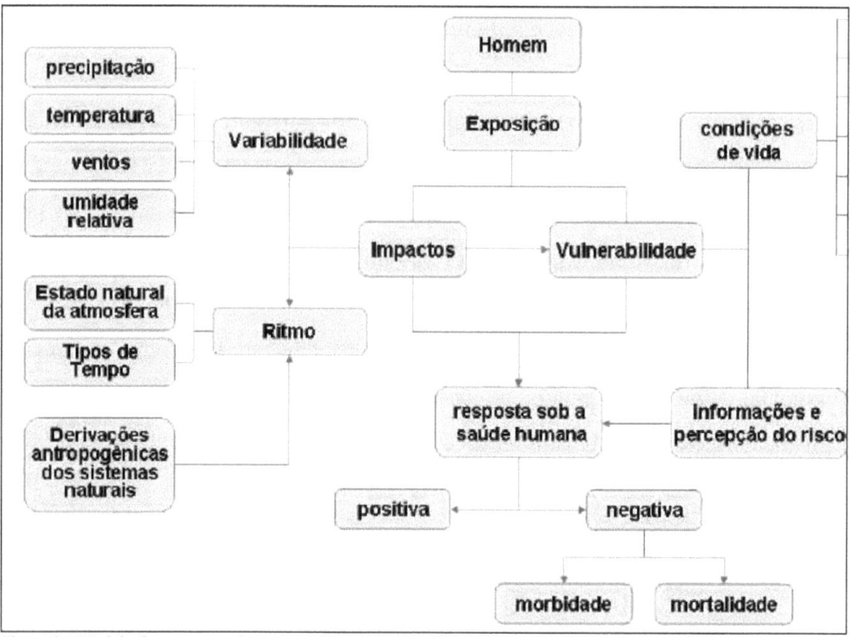

Figure 2 - Model of exposure to impacts and socio-environmental vulnerability experienced by humans
Source: CONFALONIERI, 2003

According to Monteiro (1997)

> The impact on health, especially the aggravation of certain pathologies, generated either by the behaviour of some climatic elements or by the quality of the air, and the damage to urban dynamism caused by some extremes of precipitation, will be our vehicle for returning to the idea that, after all, we are not immune to the consequences of our actions on the environment" (MONTEIRO, 1997, p.6).

Concerned about the quality of life of the urban population (and also the environment, ), geographers have become increasingly interested in studies about the climate of cities and its interference in the daily lives of the population. Thus, according to Monteiro (2003), "air pollution, heat islands, flooding in the urban space, among other forms, are highlighted in urban climates, thus reflecting peculiarities of the city's climate, which induce risks (vulnerability) to the population" (MONTEIRO, 2003, p.24).

Abreu and Ferreira (1999) analysed the main respiratory diseases that affect the urban population of Belo Horizonte in seasonal situations. An epidemiological profile of the most frequent diseases was drawn up, verifying the direct and indirect influence of climatic conditions on the human organism.

The conclusion reached in this study was that there is a clear relationship between respiratory diseases and climatic conditions. The relationship between the number of cases of respiratory diseases that occur throughout the year and temperature is inversely proportional, i.e. as the air temperature drops, there is an increase in the number of cases of hospitalisation, concentrated mainly in the autumn and winter seasons (lower temperatures). In seasons characterised by an increase in wind speed and rainfall, there is a concomitant reduction in the number of cases of respiratory diseases, due to the dispersion and elimination of pollutants and microorganisms from the air.

In this way, knowing how the weather influences health is an important method of preventing pathologies. Thus, the relationship between atmospheric conditions and the emergence of illnesses, exemplifies the importance of Geographical and Medical Climatology and Health and Medical Geography, with the aim of planning actions in favour of improving the quality of human life.

# CHAPTER 2

# METHODOLOGY

The research took a qualitative approach and also involved two distinct and complementary stages. Initially, a bibliographical review was carried out on the subject, in which guiding concepts were extracted from academic reference works.

Field research was then carried out to administer the questionnaires, which was an important stage in carrying out qualitative analyses of the subjects/participants, drawing up graphs using Microsoft Office Excel software and also verifying the context of the data and how this issue is understood and discussed by society and health professionals.

The qualitative approach, according to Oliveira and Lima (2012, p. 17) "is aimed at understanding natural or social phenomena, being a type of bias that aims to obtain a detailed analysis of the phenomena studied based on the description, search for meanings and, above all, the detailed interpretation of the data collected in a critical and consistent manner, meanings".

Thus, the qualitative approach is based on the characteristics of action research, as it starts from the principle of investigating a certain phenomenon through practice, with reference to the interfaces carried out in society and investigations through the application of interviews and questionnaires with a total of 20 health professionals: nursing technicians (10), nurses (7) and doctors (3). The hospital units that contributed to obtaining the data were: UBS (basic health unit), Regional Hospital of Cáceres-MT, municipal emergency room and Hospital São Luiz. The technical visits made it possible to find the answers to the questionnaire from a number of samples to answer the research questions. The figure below shows the number of professionals interviewed and their distribution.

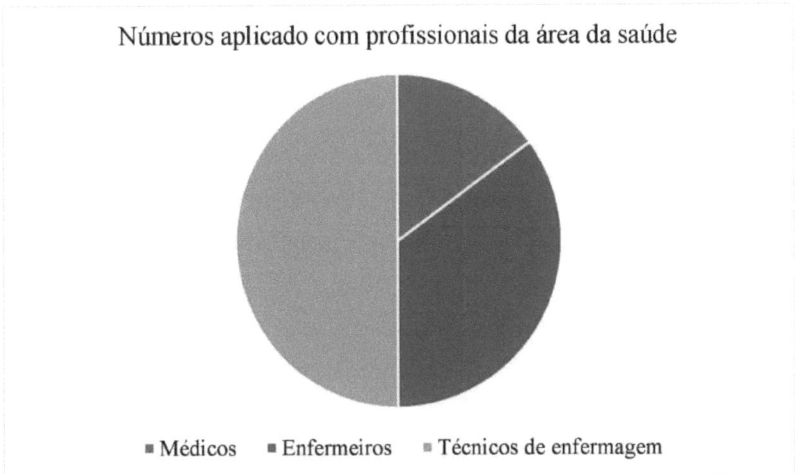


Figure 4 - Number of health professionals who took part in the interview
Prepared by: Thyago Costa Source: Field research, 2017

## 2.1. Methodological procedures and operationalisation of the research: stages, research subjects, data collection and analysis

The choice of procedures depends on the specific objectives, and therefore on the sources and materials that will be used: a bibliographical survey of texts, scientific articles and dissertations, field research, questionnaires, interviews, data collection and processing, among others. The procedures and mechanisms that were used during the development of the research corresponded to the following operational fundamentals adopted at different stages, described below.

**Stage I - Bibliographical review**

This was *a* fundamental phase for acquiring the theoretical and methodological basis of the research, which aimed to address climate issues related to health, in particular the genesis and morbidity of diseases, and consequently society's vulnerability to climatic conditions.

**Stage II - field research to conduct interviews and apply questionnaires**

The study area is located in the city of Cáceres-MT and involved both civil society (seven people living in rural areas and 13 in the urban area of the municipality of Cáceres-MT) and health professionals (nursing technicians, nurses and doctors working in the UBS basic health unit, hospital, São Luiz, regional hospital and emergency room). Data collection took place between September and December 2016, during which time it was essential to carry out three important activities described in Table 2.

Chart 2 - Activities carried out during the field research (interviews and questionnaires)

| | |
|---|---|
| **Activity I** | Preparation of questions on:<br>a) Climate and diseases;<br>b) Drought periods, atmospheric changes and diseases;<br>c) Climatic conditions and social risks ; |
| **Activity no.** | Carrying out the field research, which focused on the urban area. In this activity, questions such as the population's perception of climatic conditions and the community's own (prior) information and knowledge were fundamental to achieving the research results; |
| **Activity 111** | Visits to hospital units to verify the most common diseases and the amount of morbidity during the drought period in 2016, thus demonstrating the influence of climatic conditions on the quality and socio-environmental risk of the population. During the visit to one of the hospital units, the following questions were investigated:<br>a) When are the most cases of morbidity occurring?<br>b) What factors influence the onset of morbidities?<br>c) What are the most common illnesses related to the dry season? |

Prepared by: Thyago Costa, 2016

The subjects who took part in the research correspond to a sample of 20 people, a figure that correlates to 0.02 per cent of the total population of Cáceres and 20 professionals who work in different health units.

The main data collection tools used during the field research were: semi-structured questionnaires (Box 3 and 4), containing open and closed questions. The aim of this procedure is to analyse the issue of climate and the incidence of illness within the scope of the professional practice of health workers and also in conjunction with knowledge, which involves the population's knowledge and perception of climate on this subject.

The comparison between the two questionnaires will make it possible to see whether the answers given by the security agents are in line with the answers given by society. It will also make it possible to draw up graphs, which consist of a cartographic representation showing the percentage variation, behaviour and distribution of the samples.

Chart 3 - Questions raised during interviews with the population of Cacerde

| 1. At what time of year do the most cases of weather-related respiratory illnesses occur? ( ) Autumn ( ) Winter ( ) Spring ( ) Summer |
|---|
| 2. And why do you think some illnesses occur, especially during this period? |
| 3. What are the most common diseases related to the dry season and the most common diseases that occur during the rainy season? |
| 4. The dry season is a time of atmospheric changes and, above all, it is the time when there are severe temperature variations between the maximum and minimum, causing some health problems. What symptoms or health problems do you or someone you know experience when there is a sudden change in temperature or weather? |
| 5. What do you usually do to minimise the effects/consequences of climate drought or atmospheric changes on your daily practices? |

Chart 4 - Questions raised during interviews with health professionals - SUS

| 1. At what time of year do the most cases of weather-related illnesses occur? ( ) Autumn ( ) Winter ( ) Spring ( ) Summer |
|---|
| 2. In which months of the year, for example, are there the most cases of illness related to weather conditions or changes? And why do they occur? |
| 3. What are the most common diseases related to the dry season and the most common during the rainy season? And why do they occur in these different periods? |
| 4. The dry season happens every year and we see in the media/newspapers the number of cases of morbidity due to climatic conditions, i.e. in the dry season a large part of the population is at risk. What factors contribute to the emergence of respiratory diseases or illnesses that are common in climatic conditions? |
| 5. The dry season is a time of atmospheric changes and, above all, it is the time when there are severe temperature fluctuations between the maximum and minimum, causing some health problems. What symptoms or health problems does a person with an intransigence to atmospheric changes begin to experience when there is, for example, a sudden change in temperature or change in the weather? |
| 6. What can be done to minimise the effects of climate drought or atmospheric changes on the population's quality of life? |

**Elaboration: Thyago Costa, 2016 Source: SARTORI (2000); SOUZA (2007)**

The questions asked during the interview and the application of the questionnaires made it possible to acquire information about the elements and factors that potentiate diseases, various clarifications, an indication of the most common illnesses during the dry season, or when exceptional atmospheric changes occur, and finally, the identification of the periods with the highest incidence of diseases in Cáceres.The questionnaires were structured with open and semi-open questions and were administered by the researcher. According to Sartori (2000, p. 156), "the presence of the researcher favours clarification of the questions and encouragement to answer them".

The conversations held in the health units (regional hospital, UBS, emergency room , São Luiz hospital) and the interviews and questionnaires used contributed to the systematisation of the research results.

# CHAPTER 3

# LOCATION AND CLIMATIC CHARACTERISATION OF CÁCERES-MT: GENERAL ASPECTS

The city of Cáceres is located in the south-western region of the state of Mato Grosso. The municipal centre is 215 km from the state capital Cuiabá (Figure 3). According to estimates by the Brazilian Institute of Geography and Statistics (IBGE, 2016), it has a total population of 90,881 inhabitants. The study area is located on the left bank of the River Paraguay, and its main access route is the BR-070 federal motorway. IBGE (2016).

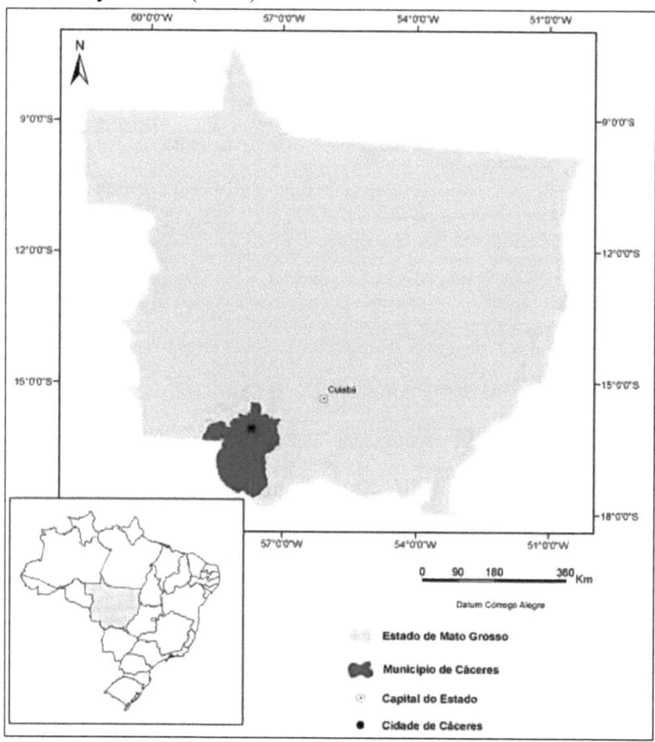


Figure 3 - Location map of the study area - Cáceres-MT
Organisation: Thyago Costa, 2016

## 3.1 Atmospheric Circulation: marks of subhumid continental tropicality

In order to better understand the climate of the study area, it is necessary to deal with atmospheric circulation, which allows us to understand weather conditions, in particular the distribution of rainfall totals, since their occurrence depends on regional atmospheric circulation conditions.

According to Nimer (1979), atmospheric mechanisms (mainly the circulation of air masses) mean that regional thermal diversification occurs as a function of geographical factors such as relief, latitude and longitude (continentality).

The author states that "all static climatic factors, such as relief, act on the climate of a given region in interaction with regional atmospheric circulation systems" (NIMER, 1979 p. 393). This demonstrates the importance of knowing the circulation systems that act on a region throughout the year in order to understand the climatic dynamics of a given area.

In addition, the study of atmospheric circulation provides a better understanding of regional climate dynamics, also taking into account geographical climate factors such as latitude, relief, vegetation and water bodies that complement the explanation of different climate patterns and even local differences between one place and another.

In this way, the elements and factors are of primary importance in climate characterisation and even make it possible to understand the dynamic behaviour of the climate in the area under study, due to the complexity of the systems that act regionally.

As for the synoptic framework, according to CPTEC (2016) the meteorological systems responsible for the climatic conditions in the area are the Continental Equatorial air masses (mEc), the Atlantic Polar Mass (mPa), the Continental Tropical Mass (mTc) and the South Atlantic Convergence Zone (ZCAS).

In regional terms, the climate of the area under study is characterised by a wide range of differences, revealed mainly by the spatial distribution of rainfall,

of the atmospheric circulation systems responsible for the rainfall regime and instability in the region, two (mEc and ZCAS) operate more frequently, while the other two atmospheric systems (mTc and mPa) operate to a limited extent, causing exceptional conditions in the region between the months of April and September. In this way, we can see that the climate in the region acts in a complex way and is the result of the combination of various atmospheric mechanisms to which geographical factors are superimposed.

## 3.2 Regional climatic characteristics of Cáceres-MT

Regionally, the area under study has the characteristics of a tropical type of articulated continentality, with a long rainy period during the year, opposed to another, drier period. Given these characteristics, the climate in the region is tropical continental, semi-humid, falling into the *Koppen* classification of the Aw type (Table 5).

Table 5 - Climate classification for Cáceres-MT

| Classification system | Typology | Climatic aspects |
|---|---|---|
| Koppen | Aw climate, megathermic, with moderate water deficiency in winter (4 to 5 months). | Tropical with summer rains and winter droughts and average temperatures ranging from 17°C (minimum) to 40°C (maximum). In the driest period there is at least one month with rainfall of less than 50 mm (winter). |

Source: ***KOPPEN***, 1928

Thus, the geographical location gives the municipality of Cáceres the climatic type identified as a tropical continental climate that is alternately humid and dry. According to Maitelli (2005), the climate of this unit can be characterised by the continentality factor, where the climatic control exerted by the relief becomes very important. Rainfall totals vary between 1,200 and 1,500 mm per year (Figure 4).

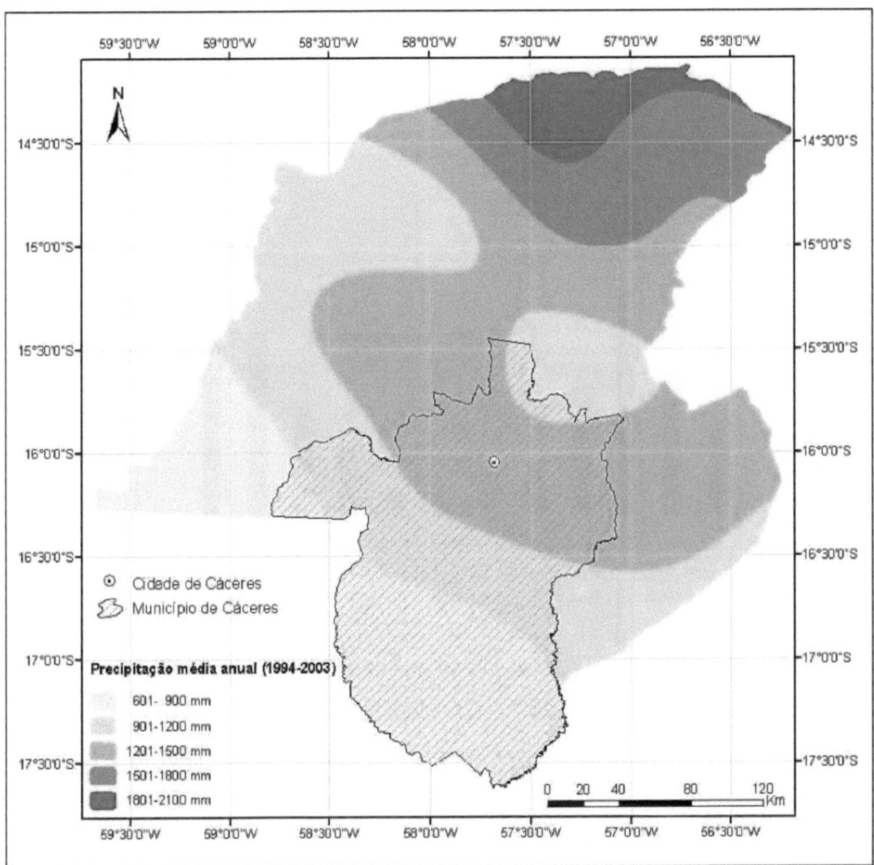


Figure 4 - Map of rainfall distribution in the Upper Paraguay River Basin - showing the municipality of Cáceres
Source: National Water Agency (ANA, 2016)

The climatic characteristics of the study area are defined by rainfall variability during the rainy season, with days of heavy rain interspersed with short periods of drought. These dry periods are known as "veranicos". According to Assad (2001), a "veranico" is a climatic event characterised by the absence of rain over a prolonged period, ranging from five to ten or twenty days, or even longer.

The rainfall pattern has the following configurations: a) the period with the highest rainfall runs from November to March, which are the five months in the study area; b) the period with the lowest rainfall comprises the months from May to September, which can be considered the five driest months; c) the three driest months within the rainfall chart are June, July and August, since the months of

May and September usually show very little rainfall during the period recorded (Figure 5).

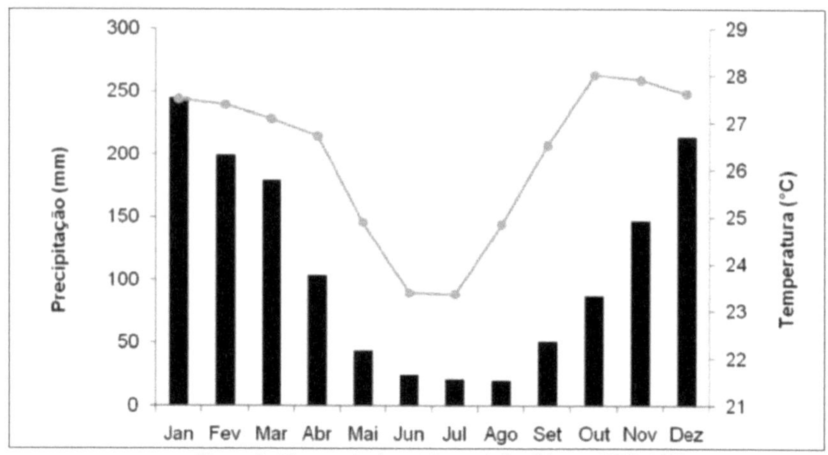


**Figure 5 - Thermopluviometric graph for Cáceres (1961-1990)**
**Source: Climatological Normals - INMET (1961 to 1990)**

Analysing the climatogram shown in figure 5, it can be seen that the highest average temperatures occur in the wet season and the lowest in the dry season, configuring the local climate in two seasons defined by the spatial and temporal distribution of rainfall. The analysis of the climatogram shows a variability in rainfall over the period analysed. It can be seen that rainfall in this region is higher in the months associated with the entry of more solar radiation into the Southern Hemisphere, i.e. the action of mesoscale convective systems, in this case the South Atlantic Convergence Zone (SACZ), which causes convective rainfall in much of Brazil and also affects this area.

In Cáceres, average annual temperatures fluctuate between 25°C and 26°C, while maximum temperatures often exceed 35°C almost all year round and during the dry season, due to the entry and movement of cold fronts and the action of the MPa, minimum temperatures reach values between 11°C and 19°C (INMET, 2016). There are meteorological records of the occurrence of "extreme" events, the lowest temperature recorded in the city was -3.5°C (1975) with a severe frost and its highest 41.8°C in 1998. Frosts are rare, with one recorded every 5 years.

The average relative humidity is 78.50%, but over the course of the year, humidity values can vary between 30% and 80%. Thus, the climatic conditions of the area under study are characterised by variations in rainfall and significant fluctuations in temperature. The combination of these elements causes thermal discomfort, which interferes with the lives of the population and, consequently, with the occurrence and potentialisation of illnesses.

# CHAPTER 4

# ATMOSPHERIC CHANGES AND THE OCCURRENCE OF DISEASES

According to the survey of health professionals, around 70 per cent of those interviewed said that the winter season (the dry season) is the time of year when people suffer most from illnesses (rhinitis, asthma, nosebleeds, coughs, allergies, breathing problems, among others). Another 30% of respondents said that summer was the season where most illnesses arise due to the weather conditions (hot and humid), and occurring in months with high rainfall, thus giving rise to illnesses such as: diarrhoea, bronchitis, dengue fever, colds, Zika virus, chikungunya, goose bumps and flu.

Based on the information provided by health professionals and in accordance with Tortura (2000), two summary tables were drawn up (Tables 6 and 7), which describe in detail the most common diseases in the rainy season and the dry season.

Table 6 - Most common diseases in the rainy season

| | |
|---|---|
| **Diarrhoea** | Diarrhoea is a very common illness that consists of frequent and uncontrolled liquid stools. It can be acute or chronic, depending on how long the symptoms last. |
| **Dengue fever** | Dengue is an acute febrile illness caused by a virus and is one of the main public health problems in the world. It is transmitted by the Aedes aegypti mosquito, which thrives in tropical and subtropical areas. |
| **Flu** | The flu is an infectious disease caused by a virus, the influenza virus (an RNA virus from the orthomyxovirus family). Flu can affect the respiratory tract (nose, larynx, pharynx, lungs, middle ear). The flu virus can attack the upper respiratory tract (throat, nose) and the lower respiratory tract (lungs) at the same time. |
| **Cold** | Hundreds of viruses are responsible for the common cold. A group of viruses called rhinoviruses accounts for around 40 per cent of all colds in adults. Typical symptoms include sneezing, excessive nasal discharge and congestion. There is often no fever. Influenza is also caused by a virus. Its symptoms include chills, fever of 38°, headache and muscle aches. |
| **Geca** | Gastroenteritis is an acute inflammation that affects the organs of the gastrointestinal system. It is characterised by increased frequency, quantity or decreased consistency of bowel movements, with or without gastric involvement (nausea and vomiting) lasting less than two weeks. |

Prepared by: Thyago da Costa (2016). Source: SUS (2016); TORTURE (2000)

Table 7 - the most common diseases during the dry season

| | |
|---|---|
| **Common Cold and Inflhienza** | Hundreds of viruses are responsible for the *common cold.* A group of viruses called rhinoviruses accounts for around 40 per cent of all colds in adults. Typical symptoms include sneezing, excessive nasal discharge and congestion. There is often no fever. Also is caused by other viruses. Its symptoms include chills, fever (38°C), headache and muscle pain. |
| **Respiratory failure** | Respiratory failure refers to a condition in which the respiratory system cannot supply enough oxygen to maintain metabolism or cannot eliminate enough carbon dioxide to prevent respiratory acidosis (a pH lower than normal). Respiratory failure always causes concomitant dysfunction in other organs. |
| **Tuberculosis (TB)** | The bacterium *Mycobacterivm tubeaculosis* produces an infectious, transmissible disease called tuberculosis (TB). TB most commonly affects the lungs and pleurae. The bacterium destroys parts of the lung tissue and this is replaced by fibrous connective tissue. |

| | |
|---|---|
| **Pneumonia** | This refers to an acute infection or inflammation of the alveoli. The alveolar saccules fill with fluid and dead leucocytes, reducing the amount of air space in the lungs. |
| **Emphysema** | In emphysema, the alveolar walls lose their elasticity and remain full of air during exhalation. When more alveoli are damaged, the lungs become permanently inflated as they lose their elasticity. To adjust to the increased size of the lung, the size of the rib cage increases, resulting in a "barrel chest". 0 Emphysema is usually caused by prolonged irritation. Cigarette smoke, air pollution and occupational exposure to industrial dust are the most common irritants. |
| **Bronchitis** | Bronchitis is inflammation of the bronchi, characterised by an increase in the glands and mucus-producing cells that line the bronchial airways. Cigarette smoke is the main cause of chronic bronchitis, i.e. bronchitis that lasts for at least three months of the year, for two successive years. |
| **Asthma** | Diseases such as asthma, bronchitis and emphysema have some degree of airway obstruction in common. The term *chronic obstructive pulmonary disease (COPD)* is used to refer to these disorders. Symptoms that can indicate significant airflow obstruction include coughing, wheezing and dyspnoea (painful or forced breathing). Asthma is a reaction, often allergic, characterised by attacks of wheezing and difficult breathing. The attacks are triggered by smooth muscle spasms in the walls of the smaller bronchi and bronchioles, causing the airways to partially close. |

Organisation: Thyago Costa, 2016 Source: SUS (2016); TORTURE (2000)

According to one of the doctors interviewed: "one of the main reasons for the genesis of illnesses during the dry season is associated with the low humidity of the air, as well as the sudden change in temperature". This understanding is also emphasised by one of the nurses interviewed: "the incidence of illnesses during the dry season occurs mainly when there are climatic variations", but anthropogenic intervention, such as fires and pollution, are contributors to the increase in illnesses.

One of the nurses warned about the rainy season, because rains and floods are extremely dangerous for health, due to the fact that

cause contamination of water accumulated on the surface. Accumulations of water, for example, can generate mosquitoes that transmit dengue fever, malaria and others that cause significant illnesses to the population. Regarding the relationship between climate and the occurrence of diseases, the nurse said that it depends a lot on the region, for example, there are Brazilian locations that do not have well-defined seasons, making it difficult to identify the months with the highest occurrence of diseases. However, it is very common for respiratory and/or infectious diseases to occur during periods of sudden changes in temperature, such as September to December and December to March.

## 4.2 Interface between climate and health: interviewees' characterisation and conceptions

This topic presents the interviewees' characterisation and conceptions of the effects of the climate on the emergence and potentiation of illnesses in the city of Cáceres. The results were organised using graphs, and statistically quantified using the percentage of responses.

### 4.2.1 Health professionals: quantification of questions 1 to 6

1- What time of year (season) poses the greatest risk for the appearance of related to climatic conditions?

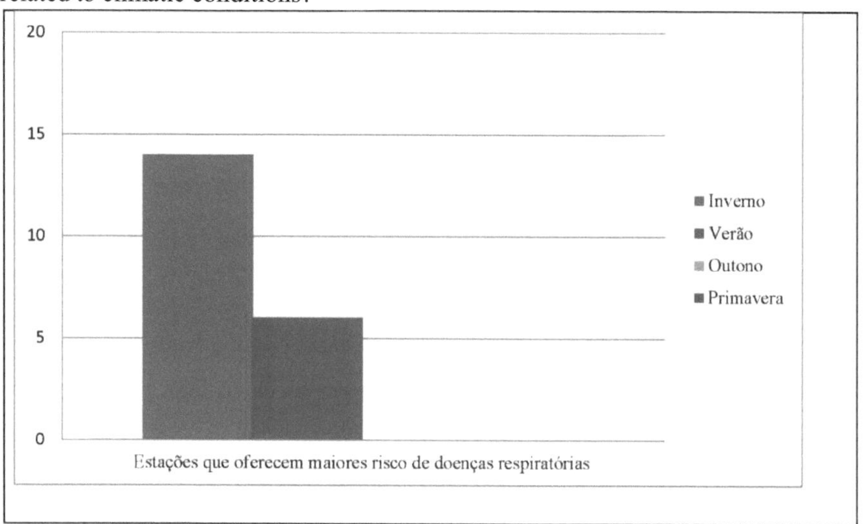


Elaboration: Thyago da Costa, 2016 Source: Field Research

2- In which months of the year do the greatest number of weather-related illnesses occur?

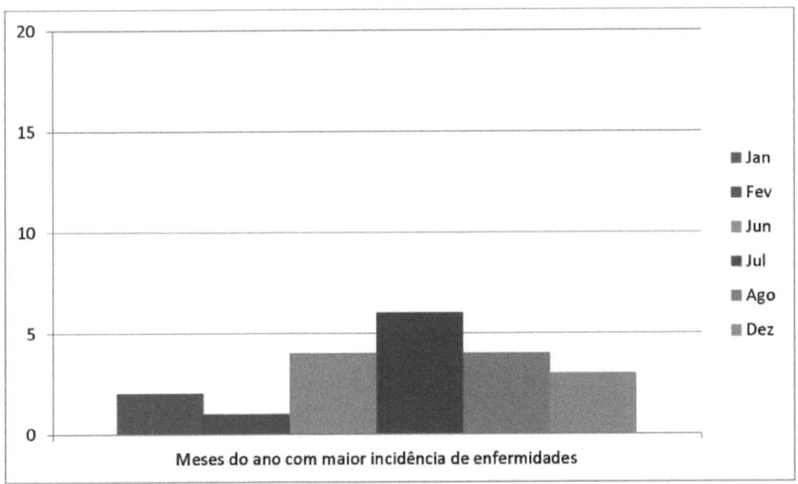

Elaboration: Thyago da Costa, 2016 Source: Field Research

2. b- Why do these illnesses occur in these months of the year?

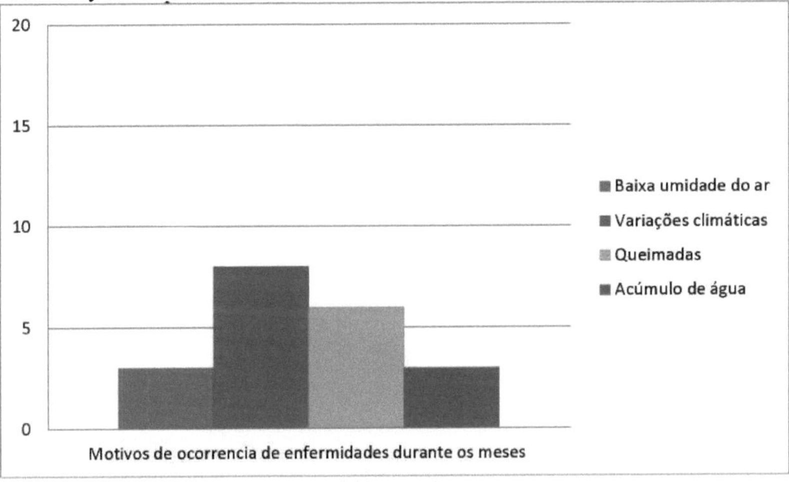

Elaboration: Thyago da Costa, 2016 Source: Field Research

3. a- The most common illnesses during the dry season?

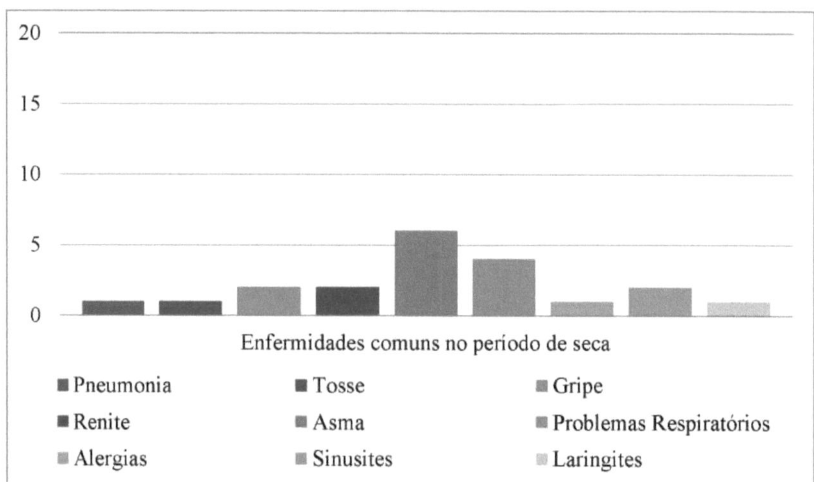


Elaboration: Thyago da Costa, 2016 Source: Field Research

3. b- The most common diseases during the rainy season.

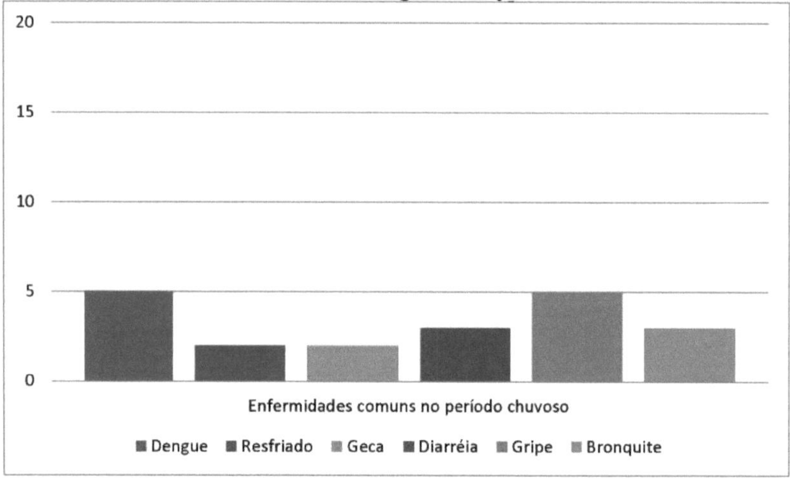


Elaboration: Thyago da Costa, 2016 Source: Field Research

4- What factors contribute to the emergence of respiratory diseases or illnesses that are common in climatic conditions?

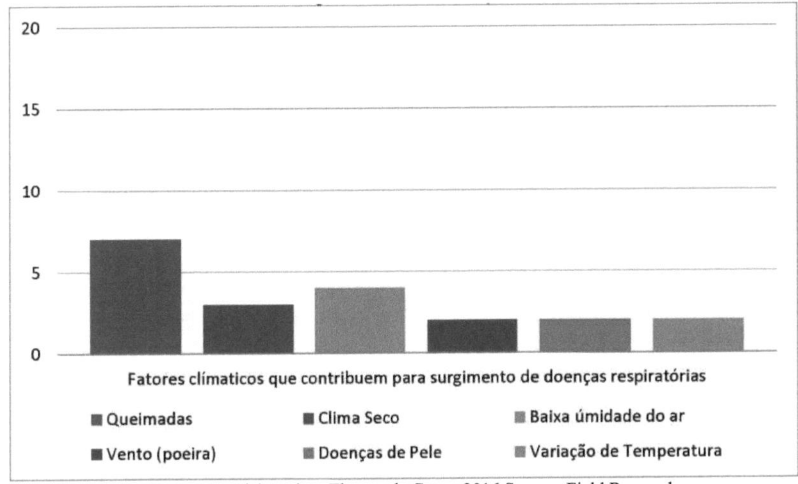


Elaboration: Thyago da Costa, 2016 Source: Field Research

5- What symptoms or health problems does a person with an intransigence to atmospheric changes begin to experience when there is, for example, a sudden rise in temperature or change in the weather?

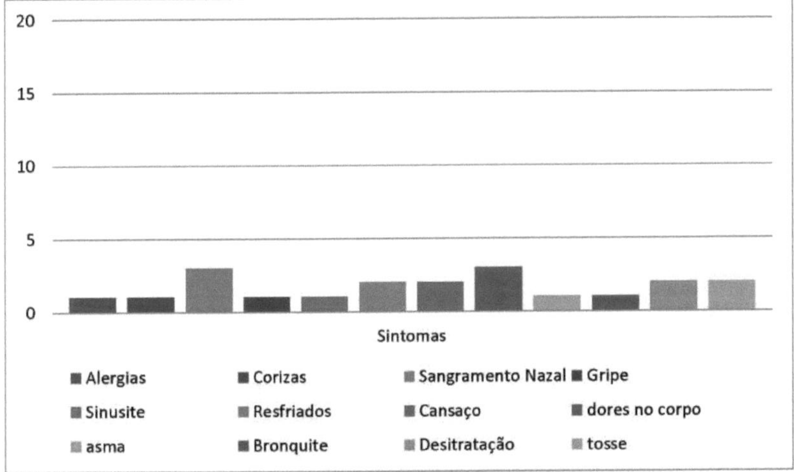


Elaboration: Thyago da Costa, 2016 Source: Field Research

6- What can be done to minimise the effects of the drought on people's quality of life?

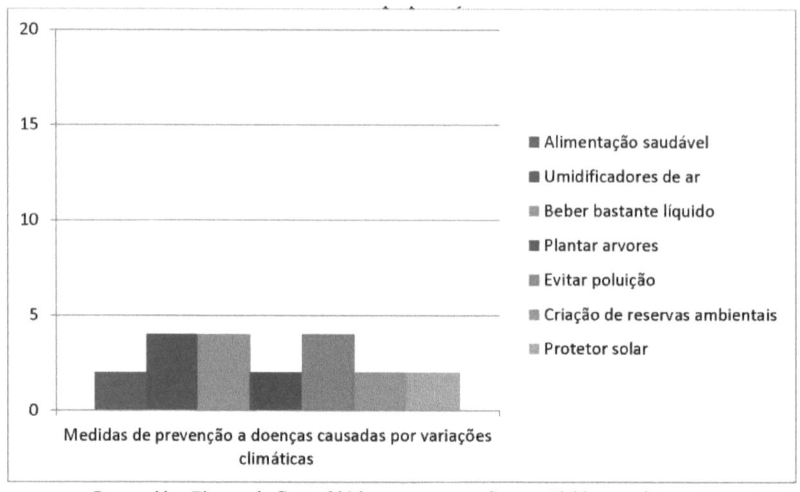


Prepared by: Thyago da Costa, 2016 Source: Field research

### 4.2.2 . Caceri community: quantification of questions 1 to 6

1. At what time of year do the most cases of weather-related respiratory illnesses occur?

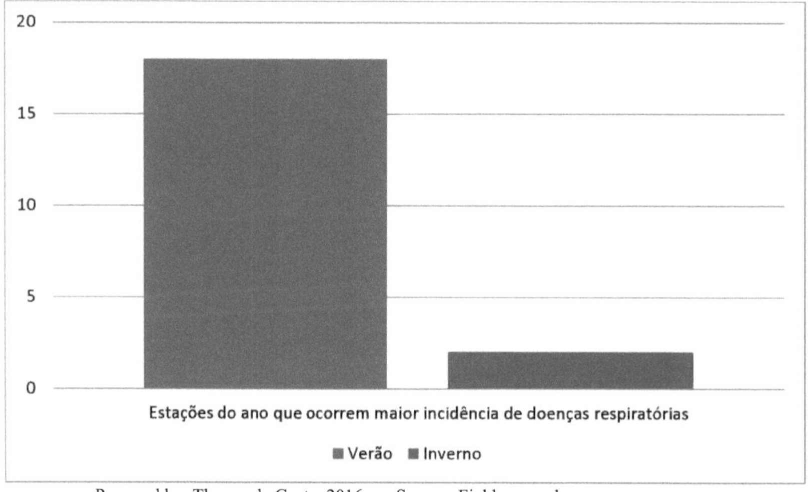


Prepared by: Thyago da Costa, 2016 Source: Field research

2. And why do you think these illnesses occur during this period?

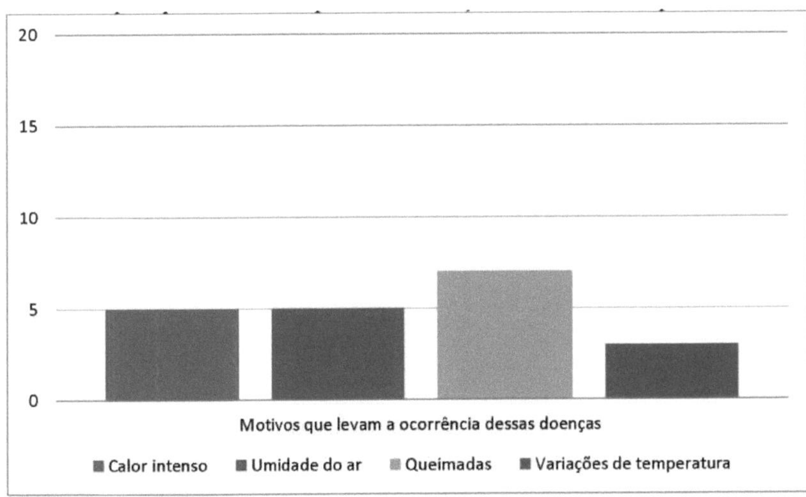


Prepared by: Thyago da Costa, 2016 Source: Field research

3. What are the most common illnesses during the dry and rainy seasons?

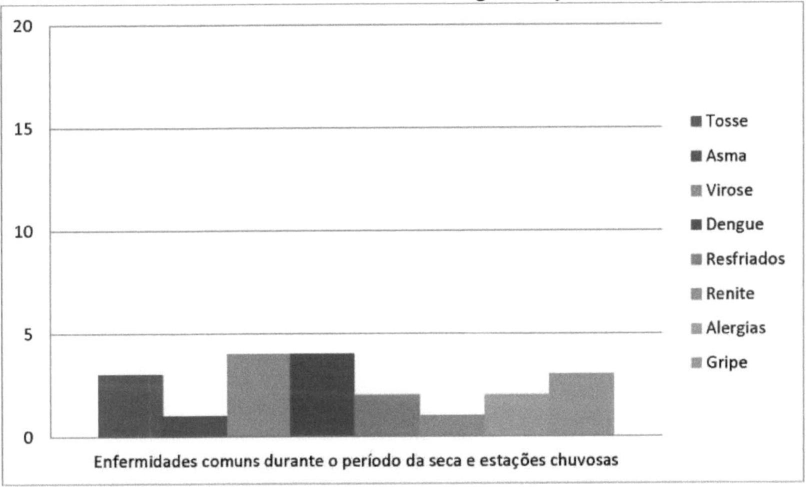


Elaboration: Thyago da Costa, 2016 Source: Field Research

4. What symptoms or health problems do you or someone you know start to feel when there is a sudden change in temperature or a change in the weather? weather?

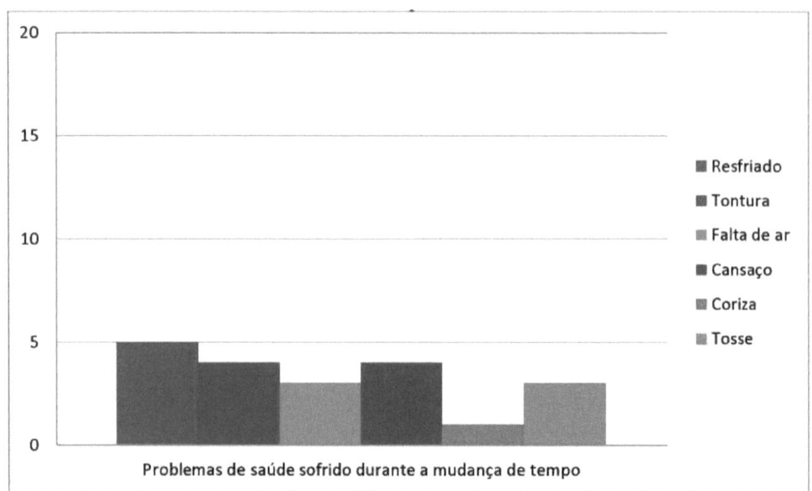


Elaboration: Thyago da Costa, 2016 Source: Field Research

5. What procedures can be taken to minimise the effects/consequences of the climate drought period or atmospheric changes on everyday practices?

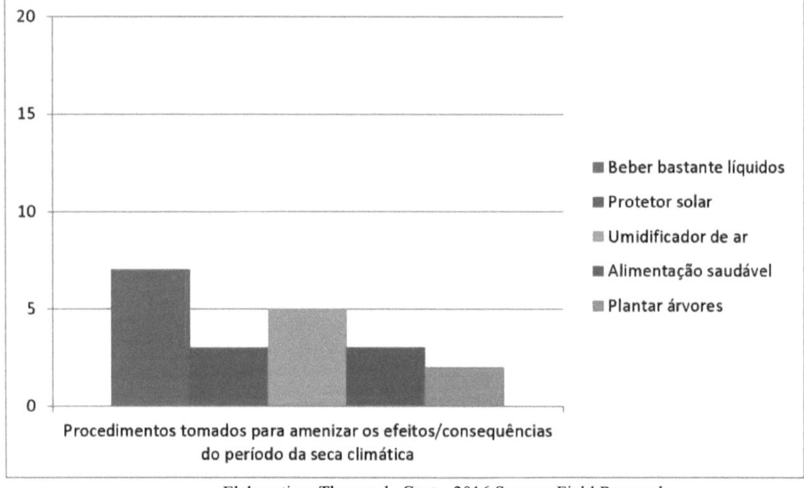


Elaboration: Thyago da Costa, 2016 Source: Field Research

From the analysis obtained through the questionnaires, the information points to the possible harmful effects of climatic elements on clinical health conditions. It can be seen that cyclical atmospheric changes influence the biological rhythm of human beings, interfering with their activities and functions.

The most common illnesses during the dry season, according to 30% of health specialists , are asthma, and during the rainy season the most common illnesses are the flu and dengue fever, according to around 25% each. The most common symptoms or health problems caused by atmospheric change are nosebleeds and body aches 15%, tiredness, coughs, dehydration, colds 10%, allergies, runny nose, bronchitis, sinusitis, asthma and flu 5%, according to the health experts.

In general, the graphs show that the most worrying seasons are winter and summer and the period when the most diseases occur due to climatic variations, low air humidity, the accumulation of water on the surface and fires.

## 4.3 Effects of climatic conditions and atmospheric changes on the incidence of diseases

In Cáceres, the distribution of rainfall has two well-defined and distinct periods, namely summer, which corresponds to December, January, February and March, being the wettest months, while winter is drier, characterised by low rainfall rates, with the months of May, June, July and August standing out.

During the summer season, which has characteristics of high and well-distributed rainfall, this is due to the high temperature level, as water evaporation intensifies due to the heat. With a different perspective, the winter season is found in the months of June, July and August, when the temperatures are lower. During this season, the lack of rainfall and the dry humidity make the temperature low, making human health vulnerable.

Rainfall, for example, plays a fundamental and very important role, "washing" the atmosphere and considerably reducing the levels of contaminants, especially suspended particulate matter, since pollution occurs in a cumulative process (SOUZA, 2007).

As these conditions worsen in winter (drought), there are also large quantities of particles suspended in the air, both from burning fossil fuels and from vehicles, fires and deforestation.

Thus, it is possible to observe that meteorological elements contribute to respiratory illnesses, but it is important to emphasise that each individual has singularities in their living conditions, such as the type of housing, social and economic situation, age group, among others, which also interfere in the increase in cases of illness.

One of the devices in the human body that has the greatest relationship with the environment is the respiratory system. Given the large amount of air that the human being

any change in the composition of the air, or even in its physical properties (such as temperature and humidity), can constitute a real problem for the individual (SOUZA, 2007).

At the municipal (sectoral) level, the highest number of hospital admissions (São Luiz and regional hospitals) for respiratory diseases occurred in the early autumn and late winter months (between April and September), when minimum temperatures dropped and droughts and lack of rainfall increased (Figure 6). During this same period, it is notable that there were higher concentrations of fire outbreaks.

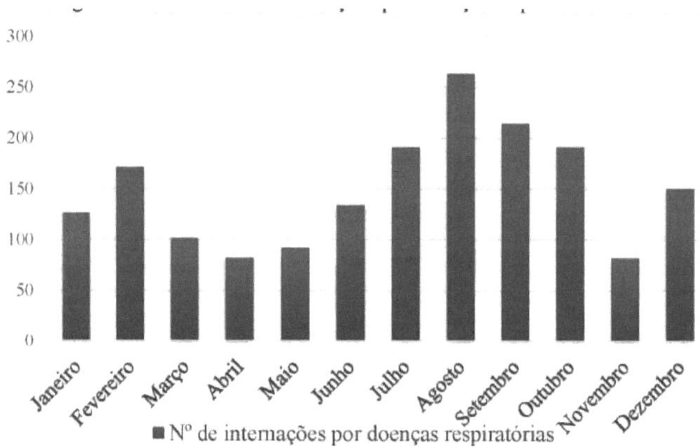


Figure 6 - Graph of the number of hospitalisations for respiratory diseases per month
Elaboration: Thyago Costa, 2016 Source: Field Research, 2016

The monthly variation in respiratory diseases, measured by the number of visits to the Unified Health System and represented in graphs at , showed a strong relationship between pathologies and temperature and relative humidity.

It is clear that the distribution of respiratory morbidity in the city of Cáceres is organised in different ways according to the seasons. The concentration of cases is mainly in the months that comprise the autumn season, due to the fact that it is a transitional season, marked by longer periods of drought and longer periods of rain.

sudden changes in temperature, as well as the population being vulnerable to climatic adversities (wearing warm clothes and humidifying the environment).

There is also a similar situation with the spring months. It is possible to see a big difference between the summer and winter seasons, which are opposite in terms of their climatic characteristics; a reality which, with regard to the winter season, increases the number of cases of respiratory tract morbidity (Figure 7).

The highest number of patients treated for respiratory diseases occurred in winter (June, July and August), a season when temperatures drop and rainfall is low, as well as the longest periods of drought and lack of precipitation. Other months have a higher rate of illnesses in the summer (December, January and February), due to the concentration of rainfall, which causes water contamination and the proliferation of disease-causing mosquitoes due to the accumulation of standing water (dengue, zika virus and chikungunya).

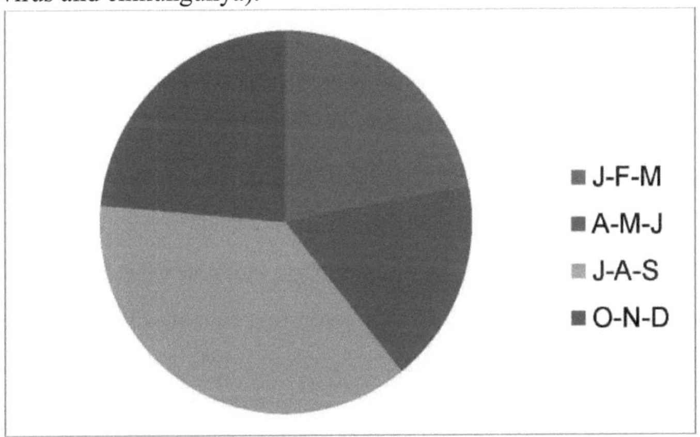


Figure 7 - Distribution graph of the number of hospitalisations
Prepared by: Thyago Costa, 2016 Source: SUS, 2013

The data shows that 22% of illnesses occur in January, February and March; 17% in April, May and June; 37% in July, August and September and 17% in October, November and December. As the climate is not solely responsible for this phenomenon, some variables that have both direct and indirect links with cases of hospitalisation due to respiratory diseases in Cáceres were circumscribed within the scope of socio-environmental quality (Table 8).

Chart 8 - Other variables that influence the onset of respiratory diseases

| | |
|---|---|
| **Vegetation** | Activities linked to deforestation or the removal of land cover make the area more prone to the effects of suspended particles, which increase the risk of respiratory tract infections; |
| **Burning** | It has great potential for concentrating respiratory diseases due to its role in the combustion of biomass, leading to the emission of PM into the atmosphere; |
| **Weather** | The circulation system of air masses and fronts is relevant to the dispersion or concentration of pollutants, mitigating or aggravating cases of disease, respectively. |

Prepared by: Thyago Costa, 2016
Source: BARROS, 2006; SOUZA, 2007; Field Research, 2016

Arbex et. *al.* (2004, p. 159) highlights the burning of biomass as a major cause of impacts and various issues related to human health:

> fire and pollution are growing problems on the planet because smoke emissions have a significant impact on the health of exposed populations. This impact includes increased mortality, hospital admissions, emergency room visits and medication use due to respiratory and cardiovascular diseases, as well as decreased lung function.

Thus, it is believed that the important factor in the greater severity of certain diseases is the sum of all the climatic variables, which are related to each other, but the socio-environmental situation has a significant influence. Thus, it can be seen that exceptional climatic conditions contribute to health problems, but each individual has unique living conditions and different levels of socio-environmental vulnerability.

Health is directly linked to the environment (understood through the interaction between society and nature, in an inseparable way), since the conditions and/or changes in the natural environment are only important to man when they are perceived or affect his well-being and way of life.

In a more detailed analysis, this research activity became necessary because it was possible to search for rich and valuable information about the relationship between extreme weather events and their aggravating factors and illnesses.

With regard to respiratory problems affecting the city of Cáceres-MT, experts point to higher numbers of patients in the months of June, July and August, which correspond to the winter season, but another season indicated in the graphs pointed out by experts.

Summer is the second season in which the most illnesses occur, corresponding to the months of December, January and February. Its characteristics are high temperatures, but the biggest concerns are the rains that occur during this season, causing flooding in certain areas and contaminating rivers and lakes. The most common illnesses during this period are diarrhoea and dengue fever.

Another important analysis is the accumulation of patients during these seasons due to the fact that neighbouring towns don't have adequate hospitals, so they end up being transferred to the health units located in Cáceres-MT. Hospitals are therefore overloaded during this period. The way to minimise this accumulation of patients are the UBS (basic health units) that are being built in order to serve all the neighbourhoods that are located near it.

# CHAPTER 5

# FINAL CONSIDERATIONS

Understanding climatic characteristics was fundamental to determining the risk of disease in Cáceres-MT. The description of the population's health conditions, together with the investigation of climatic factors that determine diseases and the evaluation of the impact of actions to change the health situation, can contribute to the evaluation of the quality of life and the work carried out in public health.

The importance of these surveys is fundamental to understanding the climatic seasons and their relationship with illnesses, but social activities also increase the potential for some types of illness. In this sense, the search for relationships between climate and health needs to be stimulated.

For a quality of life that is favourable to society, it is necessary to be within acceptable thermal comfort standards, involving climatic aspects (air temperature, humidity, wind), biological aspects (the body's response to the environment) and psychological aspects (satisfaction/dissatisfaction with external climatic conditions), presupposing an energy balance between man and the environment. Measures are also needed to improve and extend the quality of life of the inhabitant, whatever their living environment.

It is worth noting that in the health units, it was found that there is no data available at a daily level on care and deaths related to respiratory diseases, as these are not part of the group of notifiable diseases, given that there is no epidemiological surveillance of these cases because there are no outbreaks of these diseases.

In view of the aspects presented in this research and considering the climatic characteristics of the sub-humid tropical type, in order to carry out analyses that allow for intervention in the study area, it is recommended:

a) to study the short- and long-term effects by comparing morbidity rates in different areas;

b) a table of precautions to avoid respiratory infections, drawn up by pulmonologists.

c) appropriate clothing, avoiding closed and air-conditioned environments (which reduce humidity), playing sports and avoiding places with a high concentration of dust mites and fungi (for example, carpets, cushions and blankets stored for a long period of time).

d) Healthy eating strengthens the immune system;

e) drink plenty of fluids to avoid dehydration;

f) avoid exposure to the sun without sunscreen and warm clothing.

As a preventative measure, health professionals pointed out some precautions to minimise and avoid the diseases and infections caused by both the weather and the socio-environmental problem:

a) eat well so that your immune system is always "on the up", drink plenty of fluids because of dehydration due to the high temperatures, humidify the air in order to improve the environment and make breathing easier, use sunscreen and don't expose yourself to the sun without being well warmed up.

b) on socio-environmental issues, avoiding pollution in rivers and lakes and creating nature reserves and planting trees to improve air quality and the thermal balance of the surface.

c) avoid pollution and burning during the dry season, as this is a time of low rainfall and relative humidity;

In this sense, preventive measures can minimise the illnesses caused by climatic conditions. These are simple procedures, but they are important in day-to-day life in order to avoid some types of illness/pathology, as well as possible solutions that can minimise these problems caused by climate and weather changes.

Finally, it is recommended that the city's planning, environmental and health bodies, among others, work together to create solutions to reduce both socio-economic and socio-environmental vulnerability.

In conclusion, the role of climate elements and atmospheric changes in the incidence of these diseases cannot be overlooked. From this perspective, studies that address the relationship between these two segments (climate and diseases) deserve more detail and are of great relevance to society. However, it is necessary to include both climatic elements and urban environmental aspects. It is important to emphasise the need for public policies in cities that seek the quality of life and well-being of the population.

In this sense, there is a need to think about solutions for social development and the creation of public policies, one of the best and most effective instruments for improving quality of life and well-being. Since we are living in a time when high technologies and the speed of modernity are growing every day, it is possible to take some non-technological measures to reduce the illnesses associated with climatic conditions and meteorological changes.

It is therefore believed that this research could alert the public bodies responsible for health in the area in question, so that they prepare health centres and hospitals during the drought period, paying greater attention to this season, as they receive more significant numbers of patients.

The results of the research, obtained from the literature and the questionnaire, prove the relationship between climate and respiratory diseases, which, with mild temperatures (or sudden drops), rainfall concentration and long periods of drought, corroborate the aggravation of the respiratory system, increasing cases of hospitalisation.

## REFERENCES

ABREU, M. L. D. E.; FERREIRA, C. C. D. **Climatologia Médica: um estudo das doenças respiratórias em Belo Horizonte**, 1999.

ALVARES JUNIOR. O. M,; LACAVA, C. I. V.; FERNANDES, P. S. E. **Emissões Atmosféricas**. National Industrial Apprenticeship Service - SENAI. Brasília - DN, 2002. 373p.

ASSAD, E. D. **Rainfall in the Cerrado: analysis and spatialisation**. Brasília: Embrapa Cerrados. 2001.

ARBEX, M. A.; CANÇADO J, E. D.; PEREIRA, L. A. A.; BRAGA A. L.; SALDIVA P, H. N. **Biomass burning and its effects on health. Brazilian Journal of Pneumonia**, v30, n20, 2004. Available at http://www.scielo.br/pdf/jbpneu/ v.30n2/ v30, n2 a 15.pdf. Accessed on 15 November 2016.

AYOADE J. O. **Introduction to climatology for the tropics.** 10ª ed. São Paulo: **Bertrand Brasil, 2007.**

AYOADE, J. O. **Introduction to Climatology for the Tropics.** São Paulo: Difel, 1986.

BARROS, J. R. **Types of weather and incidence of respiratory diseases**: a geographical study applied to the Federal District. 2006. Thesis - São Paulo State University. Institute of Geosciences and Exact Sciences, Rio Claro Campus. São Paulo, 2006. p 121.

BARROS, Elizabeth, D. Health councils and citizen responsibility**. Ciência e Saúde coletiva**. Rio de Janeiro, v, 3, n, 1, 1998.

BORSATO, V. da A. **Brazil's climate dynamics and air masses**. 1ed. Curitiba, PR, 2016. 184.

BRAZIL, GOVERNMENT OF. **National action programme to combat desertification and mitigate the effects of drought** - PAN-Brasil. Brasília, DF: Ministry of the Environment. Secretariat of Water Resources, 2004. 242p

CARVALHO, O.; A. **Droughts and their impacts**. Available at : <http://www.geoeconomica.com.br> <pdf>Otamar de carvalho> As secas e seus impactos> Accessed on 2 November. 2016.

CASTILHO, F. J. V. **Geographical approach to urban climate and diseases in São José do Rio Preto/SP**. Rio Claro, 2006. Dissertation (Master's in Geography) Institute of Geosciences and Exact Sciences, São Paulo State University.

CENTRE FOR CLIMATE FORECASTING AND STUDIES (CPTEC). Available at: www.cptec.inpe.gov.br. Accessed: December 2016.

CONFALONIERI, V. E. C. **Climatic Variety, Social Vulnerability and Health in Brazil.** Free Lands, São Paulo - 2003.

FAGIONATTO, S. **Perception** Perception. Available at: www.lapa.ufscar.br/portugues/perc_amb.htm. 2005.

FONSECA, V. **Climate and human health. Proceedings of the VI Brazilian Symposium on Geographical Climatology**, Aracaju, SE.2004.

IBGE- Brazilian Institute of Geography and Statistics. Available at: www.IBGE.cáceres-MT (2016).

INMET- **Climatological Normals**, 1961 to 1990, 2011.

INMET -INSTITUTO NACIONAL DE METEOROLOGIA - Available at: www.inmet.gov.br/
Monthly access: November, December 2016 and January 2017.

MARTINS, A. L. and TEIXEIRA. **Acesso à atenção à saúde no sus o psf como (estreita) porta de entrada.** Recife, 2007.

MAITELLI, G. T. Atmosphere-surface interactions: the urban climate. In: MORENO, G; Tereza Higa, T.C.S; MAITELLY, G. T. (Org) Geografia de Mato Grosso: Território, Sociedade e Ambiente. Cuiabá: Entrelinas, 2005. p.238-249

MARIANO, Z. F.; ROCHA, J. R. R.; SILVA, J. F.; PEREIRA, C. C. **Respiratory diseases and winter weather conditions in Jatai-GO**. National meeting of geographers, Porto Alegre, 2010.

MESQUITA, M. E. A. Geography of Health: a study on climate and health. In: **Proceedings of the X Meeting of Latin American Geographers**. São Paulo: USP, 2005.

MONTEIRO, C, A F. Teoria e clima urbano in. **Clima urbano**. São Paulo, 2003.

NIMER, E. Climatology of Brazil. Rio de Janeiro, RJ: IBGE. 1979. 422 p.

OLIVEIRA, S. M. **Climatic elements and hospital admissions for respiratory diseases in Uberlândia (MG): perspective and challenges in climate and health studies.** Minas Gerais, 2014.

OLIVEIRA, J, C.; LIMA, S. C. **Community mobilisation and health surveillance in aedes control and dengue prevention in the district of Martinésia, Uberlândia (MG).** Minas Gerais, 2012.

PEDELABORDE, P. Introduction à L'étude scientifique du climate. Paris: SEDES, 1991.

PITTON, S. E. and DOMINGOS, A. E. Times and diseases: effects of climatic parameters on hypertensive crises in the residents of Santa Gertrudes - SP. In. **Geographical Studies**. Rio Claro, vol. 02, n°. 01, p.75-86, 2004.

SALES, G. K.;. Atmospheric factors and their impact on human health. In: **Brazilian Symposium on Geographical Climatology.** Rondonópolis. Proceedings... Rondonópolis: ABClima, 2006. 10 p.

SARTORI, M. G. B. **Clima e Percepção** (vol. 1 and 2). Doctoral thesis. Faculty of Philosophy, Letters and Human Sciences. USP, SP, 2000. p. 30-36.

SILVA, R. E.; MENDES, P. C. **O clima e as doenças respiratórias em patrimônio/MG.** Electronic journal of geography, Minas Gerais, 2012.

SOBRAL, H. R. **Air pollution and respiratory diseases in children in the greater São Paulo area: a medical geography study**. São Paulo. 1988.

SOUZA, C. G.; SANTANA NETO, J. L. **Climatic rhythm and respiratory diseases: interactions and paradoxes**. Brazilian Journal of Climatology. 2006.

SOUZA, C. G. de. **The influence of climate rhythm on respiratory morbidity in urban environments**. Presidente Prudente. Dissertation (Master's in Geography) Faculty of Science and Technology, São Paulo State University, 2007. 179 p.

SUS -SISTEMA ÚNICO DE SAÚDE. City of Cáceres - Public and Private Health Units, 2016.

TORTURE, G. J. The respiratory system. In: **Human body:** fundamentals of anatomy and physiology. Porto Alegre: Artmed Editora, p.406-431, 2000

ZEM, J. M. **Interactions between air temperature and the incidence of respiratory diseases among children in the city of São José dos Pinhais/PR.** (Master's thesis), Federal University of Paraná. Earth Sciences Sector. Department of Geography, 2004. 172

Printed by Books on Demand GmbH, Norderstedt / Germany